I'm FIVE Times Tabler!

Bill Gillham and Mark Burgess

A Magnet Book

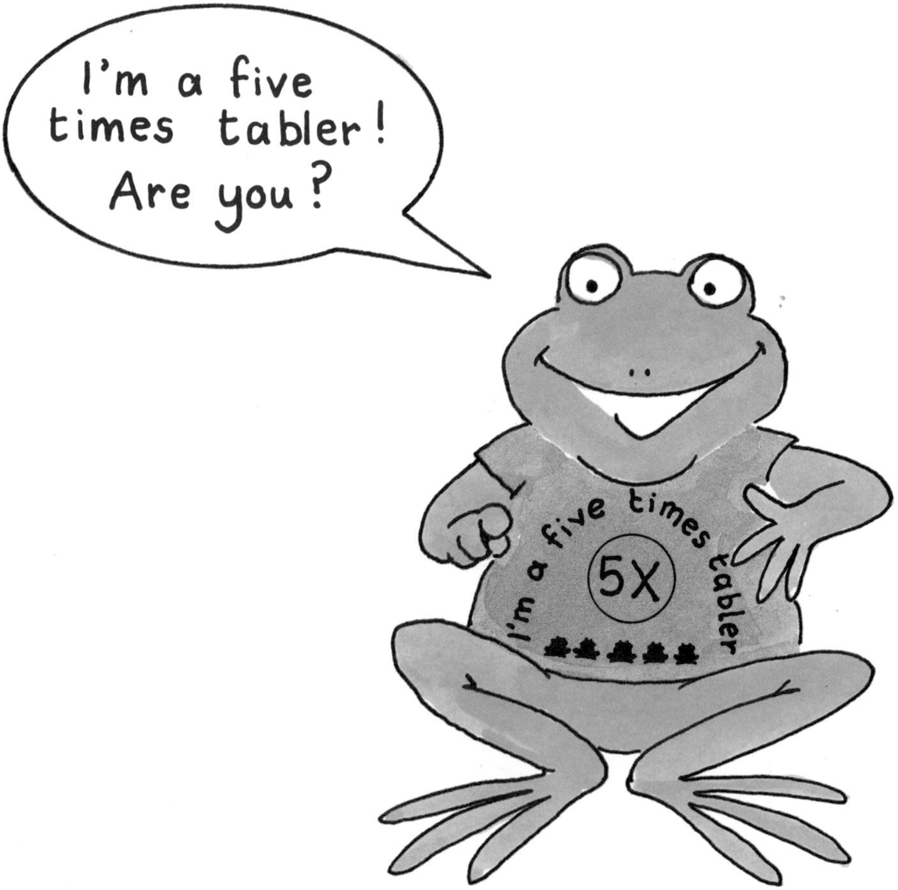
I'm a five times tabler! Are you?
I'm a five times tabler
5X

$1 \times 5 = 5$

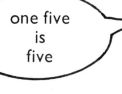

$$2 \times 5 = 10$$

$3 \times 5 = 15$

$4 \times 5 = 20$

$$5 \times 5 = 25$$

$$6 \times 5 = 30$$

$$7 \times 5 = 35$$

$$8 \times 5 = 40$$

$9 \times 5 = 45$

$$10 \times 5 = 50$$

$11 \times 5 = 55$

$12 \times 5 = 60$

$1 \times 5 = 5$

$2 \times 5 = 10$

$3 \times 5 = 15$

$4 \times 5 = 20$

$5 \times 5 = 25$

$6 \times 5 = 30$

$7 \times 5 = 35$

$8 \times 5 = 40$

$9 \times 5 = 45$

$10 \times 5 = 50$

$11 \times 5 = 55$

$12 \times 5 = 60$

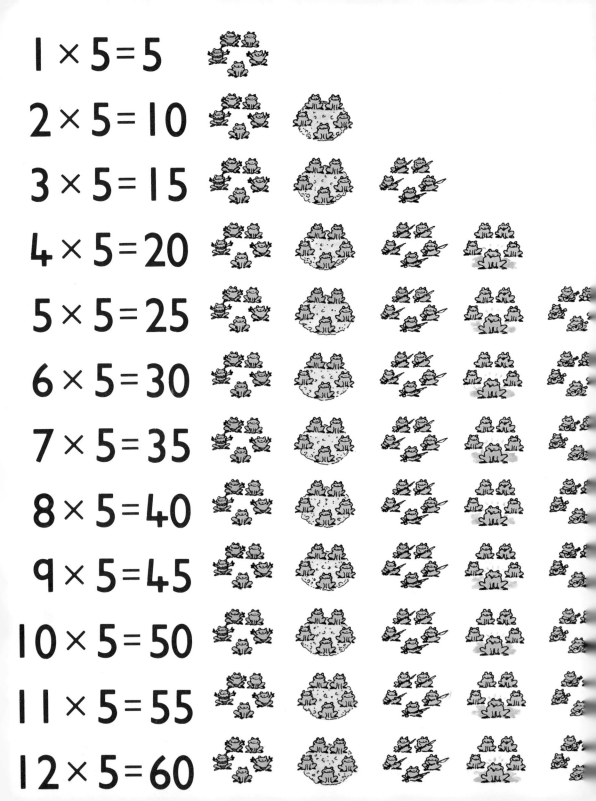

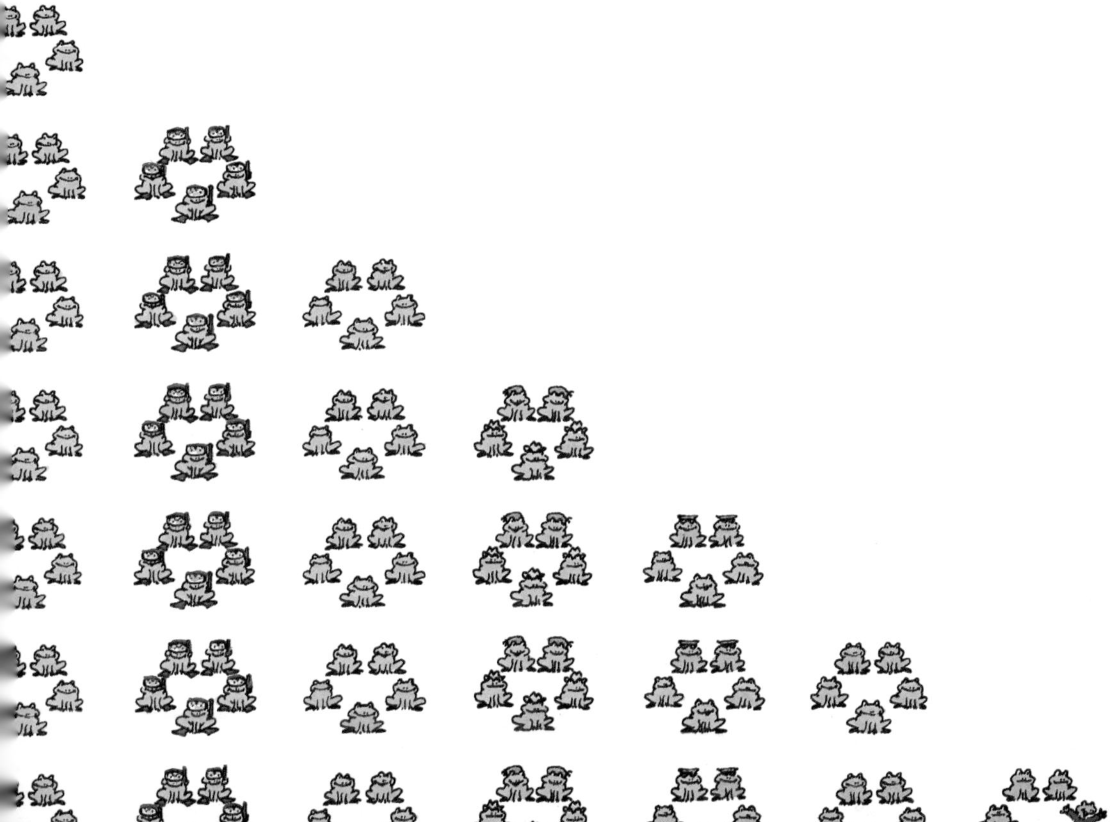

Activities

FIVE-PETAL FLOWERS

Draw a row of simple flowers each with five large and distinct petals.

Ask the child to count the petals each time and then colour them in (using a different colour for each flower). Stalks and leaves can be added but the emphasis should be on the five petals.

Let the child say the table while you write the numbers (5 – 10 – 15 etc) under each flower. The child can copy the numbers underneath.

HAND-PRINT PATTERNS

Messy but fun – and a particularly vivid
demonstration of 'fiveness'!

Sellotape six A4 sheets longways, so there are
two handprints per sheet. This gives a good
display length.

The paint needs to be gooey (poster paint is best)
and fingers should be spread apart to give
the best effect. Have a practice first.

When finished pin the pattern on the wall
and get the child to walk along saying
the table – then counting in fives.

CLOCK FACE

Remind the child that when you say the
five times table you are also *counting* in fives.

Use a 'teaching' clock face which has the minutes
marked in fives up to sixty as well as the hours
from one to twelve. Making one isn't very difficult
with card and a paper clip.

Move the minute hand from one five-minute section
to the next, saying the table as you do so.
Repeat this, with the child joining in.
Then let the child count in fives moving the clock hand.

If you have a clock with a second hand you can
repeat the activity with that. The countdown clock
between television programmes can also be used.

DOMINO PATTERN BISCUITS

The domino pattern ⁙ is one children find
attractive and is easy for them to use visually.

Making a dozen domino pattern biscuits is an
enjoyable way of focussing a child's attention
and the 'five-spot' pattern in currants
makes for easy counting in fives.

When baked twelve biscuits can be lined up
for saying the table. Then play, 'Give me two fives',
'Give me six fives', and so on – before eating!

Children need to know their tables because:
— simple multiplication, *which you can do in your head,*
is a skill of practical use in everyday life;
— the number patterns and groupings that occur
in tables help them to understand more advanced
mathematical concepts like *sets, number series*
and *progressions.*

The Times Table Books teach these ideas in a clear
and enjoyable fashion and show vividly what happen
when you multiply.

*Dr Bill Gillham is senior lecturer in the
Department of Psychology at Strathclyde University.*

First published in Great Britain in 1987
as a Magnet original
by Methuen Children's Books Ltd
11 New Fetter Lane, London EC4P 4EE
Text copyright © 1987 Bill Gillham
Illustrations copyright © 1987 Mark Burgess
Printed in Great Britain

ISBN 0 416 00232 3